Solar Energy For Beginners

A Complete Guide For Absolute Beginners

Copyright@2023

Frandre Colvin

Table Of Content

Chapter One: Introduction 5

....5

Part 1- Definition Of Solar Energy 5

Part 2- Benefits Of Utilizing Solar

Energy 6

Part 3-Brief History Of Solar Energy.....8

Chapter Two: Understanding Solar Energy 11

....11

Part 1-The Science Behind Solar Energy 11

Part 2-How Solar Panels Work 14

Part 3- Types Of Solar Panels 16

Chapter Three: Solar Energy Applications 21

....21

Part 1-Residential Solar Power 21

Part 2-Commercial Solar Power...........24

Part 4-Solar Power In Developing

Countries 27

Chapter Four: Designing And Installing

Solar Energy Systems 30

Part 1-Site Assessment........................30

Part 2-Solar Panel Selection.................32

Part 3-Inverter Selection 34

Part 4-Battery Selection 37

Part 5-Wiring And Connection 40

Chapter Five: Solar Energy Maintenance..43

Part 1-Regular Cleaning.......................43

Part 3-Battery Maintenance.................49

Chapter Six: Future Of Solar Energy.........52

....52

Part 1-Solar Energy Trends.................52

Part 2-Solar Energy Technology

Advancements 55

Part 3-Future Prospects For Solar Energy58

Chapter Seven: Conclusion.......................61

Part 1-Summary Of Key Points...........61

Part 2-Final Thoughts On Solar Energy

For Beginners 63

Chapter One: Introduction

Part 1- Definition Of Solar Energy

Solar energy refers to the energy that is
obtained from the radiation of the sun. This
energy can be harnessed and converted into
usable forms of power such as electricity,
heating, and cooling. Solar energy is a
renewable source of energy as the sun will
continue to emit radiation for billions of
years. Solar energy can be harnessed in

different ways such as through the use of solar panels or through the use of solar thermal systems.

Part 2- Benefits Of Utilizing Solar Energy

There are several benefits of using solar energy, some of which include:

1. **Renewable and Sustainable:** Solar energy is a renewable and sustainable source of energy. As long as the sun continues to shine, we will have access to this form of energy.

2. **Cost-effective:** Once installed, solar energy systems can be very cost-effective. The initial investment can be high, but the long-term savings on energy bills can be significant.

3. **Reduced carbon footprint:** Solar energy is a clean source of energy, and its use can help reduce our carbon footprint. By using solar energy, we can

reduce our reliance on fossil fuels, which contribute to greenhouse gas emissions.

4. **Energy independence:** Solar energy can provide energy independence, allowing individuals and communities to generate their own power.

5. **Low maintenance:** Solar energy systems require minimal maintenance and have a long lifespan.

6. **Versatile:** Solar energy can be used in a variety of applications, including residential, commercial, and industrial settings.

7. **Government incentives:** Many governments offer incentives for the installation of solar energy systems, making it more accessible and affordable for individuals and businesses.

Part 3-Brief History Of Solar Energy

The use of solar energy dates back
thousands of years, with early civilizations
using sunlight for warmth and light.
However, it wasn't until the 19th century
that scientists began to explore the potential
of solar energy for producing electricity.
In 1839, French physicist Alexandre-
Edmond Becquerel discovered the
photovoltaic effect, which is the ability of
certain materials to convert sunlight into
electricity. Charles Fritts, an American
inventor, invented the first solar cell in 1883
through coating selenium using a thin layer
of gold. In the early 20th century, solar
water heaters began to be used on a larger
scale, particularly in countries with abundant
sunlight such as Australia and Israel. In the
1950s, the first silicon-based solar cells were
developed, and in the 1970s, the oil crisis

prompted increased interest in solar energy as a potential alternative to fossil fuels. Since then, advances in technology have led to the development of more efficient solar cells and larger-scale solar energy systems. Today, solar energy is a rapidly growing industry, with installations around the world producing increasing amounts of clean energy.

Chapter Two: Understanding Solar Energy

Part 1-The Science Behind Solar Energy

Solar energy is based on the conversion of sunlight into usable energy. The science behind this process involves several principles of physics and materials science. When sunlight hits a solar panel, it interacts with the semiconductor material inside the panel, which typically contains layers of silicon. The energy from the sunlight is absorbed by the semiconductor material, causing some of its electrons to be excited and leave their atoms, creating a flow of electricity.

This process is known as the photovoltaic effect, which was discovered by Alexandre-Edmond Becquerel in 1839. The electricity generated by the solar panels is in the form of direct current (DC) electricity, which can then be converted into alternating current (AC) electricity through the use of an inverter.

The efficiency of a solar panel depends on several factors, including the quality of the semiconductor material, the design of the solar panel, and the amount of sunlight hitting the panel. Factors that can affect the amount of sunlight hitting the panel include weather conditions, time of day, and geographic location.

Solar energy can also be harnessed through the use of solar thermal systems, which use mirrors or lenses to focus sunlight onto a receiver, where it is converted into heat. This heat can then be used for various applications, including heating water or generating electricity.

Part 2-How Solar Panels Work

Solar panels work by converting sunlight into electricity through the photovoltaic effect.

Solar panels are made up of several layers of materials, including a top layer of protective glass, a layer of anti-reflective coating, a layer of photovoltaic cells, and a bottom layer of backsheet or protective material. The photovoltaic cells, typically made of silicon, are the heart of the solar panel. When sunlight hits the photovoltaic cells, the energy from the sunlight is absorbed by the semiconductor material, causing electrons to be excited and leave their atoms, creating a flow of electricity.

The electrical current generated by the solar panel is in the form of direct current (DC) electricity, which needs to be converted into alternating current (AC) electricity to be used in most household appliances. This is done by an inverter, which is typically installed along with the solar panel system. The amount of electricity generated by a solar panel depends on several factors, including the quality and efficiency of the photovoltaic cells, the size and orientation of the solar panel, and the amount of sunlight hitting the panel.

Overall, solar panels provide a clean and renewable source of energy, and their use is increasing rapidly around the world as more people recognize the benefits of solar power.

Part 3- Types Of Solar Panels

There are several types of solar panels available on the market, each with its own advantages and disadvantages. Below are a number of vital and popular types of solar panels:

1. **Monocrystalline solar panels:**

These are made from a single crystal of silicon, and are typically the most efficient and expensive type of solar panel. They have a uniform appearance and a high power output, but are also the most sensitive to shading and require more space.

2. **Polycrystalline solar panels:**

These are made from multiple crystals of silicon, and are typically less expensive than monocrystalline solar panels. They have a slightly lower efficiency than monocrystalline panels, but are more tolerant to shading.

3. Thin-film solar panels:

These are made from a thin layer of photovoltaic material, such as amorphous silicon or cadmium telluride. They are lightweight and flexible, and can be used in a variety of applications, but have a lower efficiency and shorter lifespan than crystalline silicon panels.

4. **Bifacial solar panels:**

These have a transparent back layer that allows sunlight to pass through and be absorbed from both sides of the panel. They can provide up to 30% more energy output than traditional solar panels.

5. Concentrated solar panels:

These use lenses or mirrors to focus sunlight onto a small area, which increases the intensity of the sunlight and produces more energy. They are normally and typically utilized in large-scale solar power plants.

The choice of solar panel type will depend on factors such as the available space, budget, efficiency, and application.

Chapter Three: Solar Energy Applications

Part 1-Residential Solar Power

Residential solar power involves the use of solar panels to generate electricity for homes. This can be a cost-effective and environmentally friendly way for homeowners to reduce their reliance on grid-supplied electricity and reduce their carbon footprint.

To install a residential solar power system, homeowners typically need to have a suitable roof or area of land that receives enough sunlight, and work with a qualified installer to determine the size and design of the system. The solar panels can be mounted on the roof or on the ground, and are typically connected to an inverter that converts the DC electricity produced by the panels into AC electricity for use in the home.

One of the main advantages of residential solar power is that it can help homeowners save money on their electricity bills. By generating their own electricity, homeowners can reduce their dependence on grid-supplied electricity and potentially earn credits for excess energy generated by their solar panels through a process called net metering.

Residential solar power can also help reduce the environmental impact of household electricity use. By generating electricity from a renewable source, homeowners can reduce their carbon footprint and contribute to efforts to combat climate change. Overall, residential solar power can be a practical and sustainable solution for homeowners looking to reduce their energy costs and environmental impact.

Part 2-Commercial Solar Power

Commercial solar power involves the use of solar panels to generate electricity for commercial buildings and businesses. This can be an effective way for businesses to reduce their energy costs, improve their sustainability, and demonstrate their commitment to corporate social responsibility.

To install a commercial solar power system, businesses typically need to work with a qualified installer to determine the size and design of the system based on their energy needs, available space, and budget. The solar panels can be mounted on the roof or on the ground, and are typically connected to an inverter that converts the DC electricity produced by the panels into AC electricity for use in the building.

One of the main advantages of commercial solar power is that it can help businesses save money on their electricity bills. By generating their own electricity, businesses can reduce their dependence on grid-supplied electricity and potentially earn credits for excess energy generated by their solar panels through a process called net metering.

Commercial solar power can also help businesses improve their sustainability and reduce their carbon footprint. By generating electricity from a renewable source, businesses can reduce their reliance on fossil fuels and contribute to efforts to combat climate change.

In addition to the economic and environmental benefits, commercial solar power can also be a powerful marketing tool for businesses. By demonstrating their commitment to sustainability and social responsibility, businesses can attract customers and differentiate themselves from competitors.

Overall, commercial solar power can be a smart investment for businesses looking to reduce their energy costs, improve their sustainability, and enhance their reputation.

Part 4-Solar Power In Developing Countries

Solar power has the potential to transform the energy landscape in developing countries, where many people lack access to reliable and affordable electricity. Solar power can be particularly useful in rural areas, where grid-connected electricity may not be available or may be prohibitively expensive to install.

In many developing countries, solar power is already being used to provide basic electricity services to households and businesses. This can include small-scale solar home systems, which typically consist of a few solar panels and a battery for storing energy, and can power lights, appliances, and mobile phones. Larger-scale solar installations can also provide electricity for schools, health clinics, and other public facilities.

One of the main advantages of solar power in developing countries is that it can help address energy poverty and improve access to basic services. By providing a reliable source of electricity, solar power can improve health outcomes, support economic development, and enhance educational opportunities.

Solar power can also be a more sustainable and cost-effective alternative to traditional fossil fuel-based energy sources in many developing countries. Unlike fossil fuels, solar power does not produce harmful emissions or contribute to climate change, and can be less expensive over the long-term due to lower operating costs and the potential for decentralized generation.

However, there are also challenges associated with expanding solar power in developing countries, including limited financing options, regulatory barriers, and lack of technical expertise. Addressing these challenges will be critical to realizing the full potential of solar power in improving access to energy and promoting sustainable development in developing countries.

Chapter Four: Designing And Installing Solar Energy Systems

Part 1-Site Assessment

Site assessment is a crucial step in the process of designing and installing a solar power system. A site assessment involves evaluating the location where the solar panels will be installed to determine its suitability for solar power generation. During a site assessment, a qualified installer will typically evaluate factors such as the orientation and slope of the roof or ground, shading from nearby trees or buildings, and the available space for installing the solar panels. The installer may also evaluate the existing electrical system and determine any upgrades that may be necessary to accommodate the solar power system.

Other factors that may be considered during a site assessment include the local climate and weather patterns, the local regulations and permitting requirements, and the availability of incentives or rebates for installing a solar power system.

By conducting a thorough site assessment, an installer can ensure that the solar power system is designed and installed to meet the specific needs and conditions of the site. This can help maximize the energy production of the system, minimize any potential issues or complications during installation, and ensure compliance with local regulations and permitting requirements.

Overall, a site assessment is an essential part of the process of designing and installing a solar power system, and can help ensure that the system is installed safely, efficiently, and effectively.

Part 2-Solar Panel Selection

Selecting the right solar panels is a critical step in designing a solar power system. When choosing solar panels, there are various aspects/factors to consider, including:

1. **Efficiency:** The efficiency of solar panels refers to the amount of sunlight they can convert into electricity. Higher efficiency solar panels can produce more electricity per square foot, which can be important for systems with limited space.

2. **Durability:** Solar panels are typically designed to last for 25 years or more, so it's important to select panels that are durable and can withstand harsh weather conditions, such as hail, wind, and snow.

3. **Cost:** Solar panels can vary significantly in price, so it's important to balance the cost of the panels with their efficiency

and durability to ensure the best return on investment.

4. **Warranty:** A strong warranty can provide peace of mind and protection in case of defects or damage to the solar panels.

5. **Brand reputation:** Choosing solar panels from a reputable brand with a proven track record can help ensure quality and reliability.

6. **Compatibility with other system components:** It's important to select solar panels that are compatible with the other components of the solar power system, such as the inverter and battery storage.

By considering these factors and working with a qualified installer, it's possible to select solar panels that are well-suited to the specific needs and conditions of a particular solar power system.

Part 3-Inverter Selection

Selecting the right inverter is an important step in designing a solar power system. The inverter is responsible for converting the DC (direct current) electricity produced by the solar panels into AC (alternating current) electricity that can be used to power appliances and electronics in the home or business.

When choosing an inverter, there are various aspects/factors to consider, including:

1. **System size:** The size of the solar power system will determine the size of the inverter needed. It's important to select an inverter that can handle the maximum power output of the solar panels.

2. **Efficiency:** The efficiency of the inverter can impact the overall energy production of the solar power system. Higher efficiency inverters can help

minimize energy losses and improve overall system performance.

3. **Type of inverter:** There are different types of inverters available, including string inverters, microinverters, and power optimizers. Each type has its own advantages and disadvantages, and the best option will depend on the specific needs and conditions of the solar power system.

4. **Monitoring and control:** Some inverters come with monitoring and control capabilities that can help track system performance and identify any issues or problems.

5. **Warranty:** A strong warranty can provide protection and peace of mind in case of defects or malfunctions in the inverter.

By considering these factors and working with a qualified installer, it's possible to select an inverter that is well-suited to the specific needs and conditions of a particular solar power system.

Part 4-Battery Selection

Battery selection is an important consideration for solar power systems that include battery storage. Batteries allow excess energy produced by the solar panels to be stored for later use, which can help maximize energy independence and reduce reliance on the grid.

When selecting batteries for a solar power system, there are several factors to consider, including:

1. **Battery type:** There are several types of batteries available, including lead-acid, lithium-ion, and flow batteries. Each type has its own advantages and disadvantages, and the best option will depend on the specific needs and conditions of the solar power system.

2. **Capacity:** The capacity of the batteries will determine how much energy can be stored and used later. It's important to select batteries with enough capacity to meet the energy needs of the home or business.

3. **Depth of discharge:** The depth of discharge refers to how much of the battery's capacity can be used before it needs to be recharged. It's important to select batteries with a deep depth of discharge to maximize their usable capacity.

4. **Cycle life:** The cycle life of a battery refers to how many times it can be charged and discharged before it begins to degrade. It's important to select batteries with a high cycle life to ensure they last for as long as possible.

5. **Cost:** Batteries can vary significantly in price, so it's important to balance the

cost of the batteries with their capacity, cycle life, and other factors to ensure the best return on investment.

6. **Warranty:** A strong warranty can provide protection and peace of mind in case of defects or malfunctions in the batteries.

By considering these factors and working with a qualified installer, it's possible to select batteries that are well-suited to the specific needs and conditions of a particular solar power system.

Part 5-Wiring And Connection

Wiring and connection are crucial aspects of a solar power system design. Proper wiring and connection ensure the safe and efficient flow of electricity throughout the system. Here are some key considerations when it comes to wiring and connection for a solar power system:

1. **Wire gauge:** The gauge of the wires used in a solar power system depends on the amount of current that will be flowing through them. Higher current systems require thicker wires with lower gauge numbers to ensure proper conductivity and minimize energy loss.

2. **Conduit:** Conduit is used to protect the wires from damage and provide a safe environment for the flow of electricity. It's important to use conduit that is

appropriate for the environment in which the solar power system will be installed.

3. **Grounding:** Proper grounding is critical for safety and system performance. The solar panels, inverter, and other system components must be grounded according to local building codes and regulations.

4. **Connectors:** High-quality connectors are important for ensuring a secure and reliable connection between the various components of the solar power system. It's important to use connectors that are appropriate for the type of wire being used.

5. **Fuse and breaker sizing:** Fuses and breakers are used to protect the system from overloading or short circuits. It's important to size the fuses and breakers appropriately for the size of the system and the maximum current that will be flowing through it.

By considering these factors and working with a qualified installer, it's possible to ensure that the wiring and connection for a solar power system are safe, efficient, and reliable.

Chapter Five: Solar Energy Maintenance

Part 1-Regular Cleaning

Regular cleaning is an important aspect of solar panel maintenance to ensure that they operate at maximum efficiency. Dust, dirt, bird droppings, and other debris can accumulate on the surface of the solar panels, reducing their ability to capture sunlight and convert it into electricity. Here are some tips for regular cleaning of solar panels:

1. **Frequency:** The frequency of cleaning will depend on the local climate and environmental conditions. In areas with heavy dust or pollen, cleaning may be required more frequently than in areas with cleaner air.

2. **Time of day:** The best time to clean solar panels is early in the morning or late in the evening when the panels are

cool to the touch. Cleaning the panels during the hottest part of the day can increase the risk of damage due to rapid temperature changes.

3. **Cleaning solution:** A mild soap solution and water are typically sufficient for cleaning solar panels. Avoid using harsh chemicals or abrasive materials that can scratch the surface of the panels.

4. **Tools:** Soft-bristled brushes or squeegees are good tools for cleaning solar panels. Avoid using pressure washers or high-pressure sprayers that can damage the panels.

5. **Safety:** When cleaning solar panels, it's important to take safety precautions to avoid injury. Use a sturdy ladder or scaffolding to reach the panels, and wear non-slip shoes and safety equipment as necessary.

By following these tips for regular cleaning, it's possible to keep solar panels operating at maximum efficiency and extend their lifespan.

Part 2-Electrical Inspection

An electrical inspection is an important aspect of solar panel maintenance to ensure that the electrical components of the system are operating safely and efficiently. Here are some key considerations for an electrical inspection of a solar power system:

1. **Checking connections:** The electrical connections between the solar panels, inverter, batteries, and other components should be checked for tightness and corrosion. Loose or corroded connections can cause energy loss and increase the risk of electrical fires.

2. **Inspecting wiring:** The wiring of the solar power system should be inspected for damage, wear, and proper sizing. Any damaged or worn wiring should be replaced to ensure safe and efficient operation.

3. **Checking the inverter:** The inverter is a critical component of a solar power system that converts DC electricity from the solar panels into AC electricity that can be used by household appliances. The inverter should be checked for proper operation and any signs of damage or wear.

4. **Testing battery performance:** If the solar power system includes batteries for energy storage, the batteries should be tested for proper performance and capacity. Any batteries that are not performing properly should be replaced.

5. **Compliance with local codes and regulations:** The solar power system should be checked for compliance with local building codes and regulations. Any non-compliant components should be replaced or modified to ensure safe and legal operation.

By conducting regular electrical inspections of a solar power system, it's possible to identify and address potential problems before they become serious issues. It's important to work with a qualified installer or electrician for electrical inspections to ensure that the system is safe and operating at maximum efficiency.

Part 3-Battery Maintenance

Battery maintenance is an important aspect of solar power system maintenance, especially if the system includes batteries for energy storage. Here are some key considerations for battery maintenance:

1. **Regular inspections:** Batteries should be inspected regularly for signs of damage or wear. Any damaged batteries should be replaced immediately to avoid potential safety hazards.

2. **Monitoring battery levels:** The charge levels of batteries should be monitored regularly to ensure that they are functioning properly. Overcharging or undercharging of batteries can reduce their lifespan and may cause safety hazards.

3. **Cleaning batteries**: Batteries should be cleaned regularly to prevent the buildup of dirt and debris. Accumulated dirt and debris can reduce the efficiency of the battery and increase the risk of damage or failure.

4. **Temperature control:** Batteries should be kept in a location with stable temperature control to prevent overheating or freezing. Extreme temperatures can reduce the lifespan of batteries and may cause safety hazards.

5. **Battery replacements:** Batteries have a limited lifespan and should be replaced when they no longer function properly. Replacement batteries should be properly matched to the system to ensure proper performance and safety.

By following these key considerations for battery maintenance, it's possible to ensure that batteries in a solar power system are functioning properly and safely. It's important to work with a qualified installer or electrician for battery maintenance to ensure that the system is properly maintained and to avoid potential safety hazards.

Chapter Six: Future Of Solar Energy

Part 1-Solar Energy Trends

Solar energy is becoming an increasingly important part of the global energy landscape, with growing adoption and investment in solar power systems. Below are a number of vital and key trends in solar energy:

1. **Falling solar panel costs:** The cost of solar panels has fallen dramatically in

recent years, making solar energy more accessible and cost-effective for homes, businesses, and governments.

2. **Increased adoption:** Solar energy adoption has been growing rapidly around the world, with many countries setting ambitious targets for renewable energy adoption.

3. **Innovations in solar technology:** Advances in solar technology are increasing the efficiency and output of solar panels, making them more cost-effective and reducing the amount of land needed for solar power systems.

4. **Energy storage:** Battery storage technology is becoming more affordable and efficient, making it possible for solar power systems to store excess energy and provide power during periods of low sunlight.

5. **Government support:** Governments around the world are providing incentives and support for solar energy adoption, including tax credits, subsidies, and regulatory frameworks that encourage renewable energy investment.

Overall, solar energy is expected to continue growing in importance as a clean and renewable energy source, providing an important solution to climate change and energy security challenges.

Part 2-Solar Energy Technology Advancements

Solar energy technology is constantly evolving and improving, leading to greater efficiency, affordability, and versatility in solar power systems. Here are a number of vital and key advancements in solar energy technology:

1. **Higher efficiency solar cells:** Advances in solar cell technology have led to cells that are more efficient at converting sunlight into electricity, reducing the amount of land needed for solar power systems.

2. **Thin-film solar cells:** Thin-film solar cells are a newer type of solar cell that use a thin layer of semiconductor material, making them more lightweight and flexible than traditional silicon solar cells.

3. **Energy storage:** Battery storage technology has improved, making it possible for solar power systems to store excess energy and provide power during periods of low sunlight.

4. **Smart inverters:** Smart inverters are able to control the output of solar power systems more efficiently, allowing them to be integrated more seamlessly into the grid.

5. **Solar trackers:** Solar trackers use sensors and motors to follow the movement of the sun, increasing the amount of sunlight that solar panels receive throughout the day and improving overall system efficiency.

Overall, these advancements in solar energy technology are leading to more cost-effective and efficient solar power systems, making solar energy an increasingly viable option for homes, businesses, and governments around the world.

Part 3-Future Prospects For Solar Energy

The future prospects for solar energy are bright, with solar energy becoming an increasingly important part of the global energy mix. Below are a number of vital and key future prospects for solar energy:

1. **Continued cost reductions:** The cost of solar energy is expected to continue falling as technology improves and economies of scale are achieved, making solar energy more accessible and affordable.

2. **Increased adoption:** Solar energy adoption is expected to continue growing rapidly around the world, as countries set ambitious targets for renewable energy adoption and solar energy becomes more cost-effective and accessible.

3. **Energy storage:** Battery storage technology is expected to continue improving, making it possible for solar power systems to store excess energy and provide power during periods of low sunlight, increasing the flexibility and reliability of solar energy systems.

4. **Smart grid integration:** As solar energy adoption grows, there will be an increased need for smart grid technology to manage the integration of solar energy into the grid, allowing for more efficient use of renewable energy resources.

5. **Solar in emerging markets:** As solar energy becomes more affordable and accessible, it is expected to play an increasingly important role in meeting energy demand in emerging markets, particularly in regions with abundant sunlight.

Overall, the future prospects for solar energy are promising, with solar energy playing an important role in addressing climate change and energy security challenges while providing clean and renewable energy to homes, businesses, and governments around the world.

Chapter Seven: Conclusion

Part 1-Summary Of Key Points

In summary, below are a number of vital and key points about solar energy:

- Solar energy is a renewable energy source that uses sunlight to generate electricity.

- Solar energy has many benefits, including reducing greenhouse gas emissions, improving energy security, and reducing dependence on fossil fuels.

- The history of solar energy dates back to the 1800s, but significant advancements have been made in recent years, including falling solar panel costs and increased adoption.

- Solar panels work by converting sunlight into electricity through the use of solar cells.

- There are several types of solar panels available, including monocrystalline, polycrystalline, and thin-film solar panels.
- Solar energy can be utilized for residential, commercial, and industrial purposes.
- Site assessment, solar panel selection, inverter selection, battery selection, and wiring and connection are all important factors to consider when installing a solar power system.
- Regular cleaning, electrical inspection, and battery maintenance are important for ensuring the optimal performance and longevity of a solar power system.
- Future prospects for solar energy are bright, with continued cost reductions, increased adoption, and advancements in energy storage and smart grid technology expected.

Part 2-Final Thoughts On Solar Energy For Beginners

Solar energy is a clean, renewable, and increasingly affordable energy source that has the potential to play a significant role in addressing climate change, improving energy security, and providing access to electricity in emerging markets. As a beginner, understanding the basics of solar energy, including how solar panels work, the different types of solar panels available, and the factors to consider when installing a solar power system, is essential.

While solar energy has many benefits, it is important to consider the unique circumstances and needs of your specific situation when deciding whether solar energy is the right choice for you. However, with the continued advancements in solar energy technology and increasing adoption around the world, it is clear that solar energy will continue to play an important role in the global energy mix. By learning about solar energy and its potential, beginners can be better equipped to make informed decisions about their energy use and contribute to a more sustainable future.